Alzheimer's
Disease is Treatable

ALZHEIMER'S DISEASE IS TREATABLE

OVERVIEW

Imagine losing the intellectual functions that you have learned throughout your life. **The loss of your functional and intellectual capacity gradually gets worse to the point that you cannot function independently in any social or other situations.** Finally, you are unable to perform normal activities of daily living. You lose human qualities such as memory, abstract thinking, language, naming, calculation, comprehension, and the ability to solve simple problems. Additionally, you have personality changes and behavioral problems. You show difficulty with speaking, dressing, eating, writing, and recognizing caregivers, family members, loved ones, and friends. This is Dementia.

Alzheimer's disease (AD) is the most common form of dementia. AD is a serious and devastating disease in which there is **irreversible loss of the all mental capabilities** of the patient. Initial symptoms are subtle, but there is a progressive decline in all aspects of memory and cognition. Memory loss in Alzheimer's disease is much more disastrous than just the simple aging process. **1 in 8 Americans above the age of 65 live with the disease.** Just like in most diseases, **early detection** will provide the most benefit in curbing disease progression and side effects. With early detection, you will get all the benefits of the current therapies available. You will also have some time to plan your future, because Alzheimer's disease is not only an expensive and progressive disease; it is ultimately fatal. AD is the **sixth leading cause of death** amongst Americans, affecting more than 5.1 million people. Joining a local Alzheimer's group or the Alzheimer's Association can help you get the information you need as well as connecting you with others who are suffering from the same disease as you.

The **prevalence** of the disease increases with age, and is **most common after age 65**. About 13% of the people over age 65, 45% over age 80, and 46% over age 85 have the disease. The average age at the time of diagnosis is 70 years.

Americans affected with Alzheimer's disease with 370,000 new cases per year. About 500,000 are suffering from early onset Alzheimer's or other dementias. There will be roughly 15 million Americans affected with Alzheimer's (AD) by the year 2050. **About 2/3 of people with Alzheimer's are women.** The process **starts decades before** the diagnosis.

<u>Alzheimer's Disease 13 Ominous Early Signs</u>

DR. ALOIS ALZHEIMER :

A German psychoneuropathologist published a paper in **1907** describing the previously undocumented condition. He was fascinated by a patient at the Frankfurt Asylum named Mrs. Auguste D., a 51-year-old female who presented with short-term memory loss and behavioral changes as she regressed. When she died in 1906, Dr. Alzheimer sent her brain to Munich where he worked with Dr. Emil Kraeplin. They identified amyloid plaques and neurofibrillary tangles which are the hallmark features of Alzheimer's disease, still required for pathological diagnosis. Although it was once thought to be a rare disorder, **it is now the most** common form of dementia. Dementia is a Latin word, in which DE means none and MENTIA means mind. It is known that 50-80% of all dementias are due to Alzheimer's disease (AD).

Although clinicians can make the diagnosis on clinical presentation, clinical findings, and a mental status exam; the autopsy and/or biopsy are the only way to make a definite diagnosis.

Stages of Alzheimer's disease / Dementia:

The process of Alzheimer's disease starts several decades before the actual diagnosis. Most experts developed several methods to categorize the pattern of the disease progression. The stages of disease progression are not used to determine life expectancy; rather they are used for educational purposes and treatment planning. It can help you plan for lifestyle changes, financial needs, and medical necessities. **The duration of the disease process is 0-20 years**. The new system was developed by Dr. Barry Reisberg, from the New York University School of Medicine and is being used in research as well as to qualify patients for hospice in later stages.

A Mini Mental Status Examination (MMSE) is done to determine the severity of the disease. MMSE is a 30-point test for dementia administered by medical professionals to help diagnose Alzheimer's, with results from an EEG (Electroencephalogram), CT Scan, and MRI being within normal limits. One can do a SLUMS test for dementia, which is more detailed compared to a MMSE.

MMSE = Mini Mental Status Exam (30-point scale)

<10 = Severe.

10 - 20 = Moderate Dementia.

20 to <25 = Mild Dementia.

25 to < 28-30 = MCI (Mild Cognitive Impairment)

The process of Alzheimer's disease starts several decades before the actual diagnosis. There is a stage called MCI (Mild Cognitive Insufficiency) in which the patient will have mild forgetfulness with an MMSE score above 26. This could be known as the Pre-Alzheimer's stage. We place them in stages based on their signs and symptoms. The staging does not fit every single patient every single day.

Stages 1-7, with detailed description of mild, moderate, and severe dementia is as follows:

1. Normal adult: no cognitive decline: normal, no functional decline.

2. Very mild cognitive decline: MMSE 29 (SLUMS) aware of minor functional decline, misplacing things, some work difficulties. Friends, relatives, and doctors are not able to detect any appreciable loss of memory.

3. MCI: Mild Cognitive Impairment: Early Dementia: **everybody around him/her starts noticing the decline**. MMSE or SLUMS 25-29, poor organizational skills, unable to travel to new locations, difficulty performing job skills, may last up to seven years, word-finding difficulties, **trouble naming objects**, a decline in abilities to plan or organize.

4. Mild Dementia: MMSE 20-23 or SLUMS mild/early stages of Alzheimer: moderate cognitive decline. May last up to two years, **needs assistance with** complicated tasks, handling of financial affairs, arranging a party, shopping, recalling of recent events, and calculating.

5. Moderately dementia: **MMSE or SLUMS 10-20, needs assistance in getting proper clothing for the occasion or season,** wants to keep the same clothing, trouble recalling place and date, still able to recall their names of spouse and children, requires no assistance with toileting/eating.

6. Moderately severe dementia: MMSE or SLUMS 0-9, lost recent recall, may recall their own name but forgets the names of spouses. The patient needs assistance in dressing, bathing, toiling, and incontinent of bowel and bladder. They will show repetitive behavior, personality changes, suspicious and delusional thoughts, and will have disrupted sleep.

7. Severe dementia: MMSE or SLUMS <10 and worsening, late stages of Alzheimer's disease: unrecognizable speech, speech <6 intelligible words, progressive loss of abilities to walk, talk, eat, dress, sit up, smile, hold the head up, and incontinence of bowel and bladder.

Mild/Early Dementia:

The first stage of Alzheimer's disease lasts roughly 3-10 years or more. The patient shows mild forgetfulness and misplaces things. Learning new concepts, words, and tasks are not only difficult, but impossible. He is unable to name friends or recognize objects. He has difficulty with recent memory, and has a tendency to become lost. He shows depression, anxiety, and personality changes. He has difficulty balancing a checkbook.

There are special scans called PET scans and SPECT scans that show decreased metabolism and blood flow to certain parts of the brain. The PET scan shows a decreased metabolism in the temporal and parietal area of the brain. These tests, however, are not done on a routine basis.

Depression is very common in AD. Depression affects >20% of patients, and 50% of caregivers. The newer antidepressants help by relieving panic, anxiety, tension, and depression. The most important thing in this stage is to find a physician who is knowledgeable, and someone the family can cooperate with. They need to work with the doctors and medical professionals to rule out any treatable causes of forgetfulness and take care of the patient in the best possible way

Treatment is available at this stage. Research suggests the medications like Aricept (donepezil), Exelon (rivastigmine), Reminyl (galantamine), or Namenda (memantine) could be prescribed. Memantine (Namenda) has been approved for moderate-to-severe dementia, since 2003. Insurance pays for Namenda only in moderate to severe dementia. In addition, Motrin (NSAIDS), aspirin, vitamin E, and the newer antidepressants may help some patients. Motrin or Naprosyn should not be used for more than 10 days, per FDA. Neurobics (brain exercises), such as crossword puzzles, playing cards, etc., do help. Please visit neurobics sites for more exercises, but only perform exercises that are not dangerous. (www.neurobics.com). _____

<u>Moderate Dementia:</u>

In this stage, the patient continues to show more forgetfulness. There is a severe loss of recent and remote memory, and he continues to decline in all brain functions.

His comprehension is poor and his language continues to get worse. He cannot maintain a checkbook, nor is he able to pay bills properly. He is disoriented as to time, place, and person. He shows poor comprehension, calculation, and social skills. He becomes increasingly confused and argumentative, which makes him easily irritable. He shows some depression, delusional thoughts, agitation, and restlessness. He is anxious and paces the floor wandering in and out of the house. He loses things or hides things then believes that imaginary people are stealing them. He has hallucinations. If you do take him somewhere, he wants to go back home. If he is in the house and starts to get confused, he will tell you he wants to go back home. These characteristics are a clear indication that the patient has problems with cognitive functions. Fire hazards increase; therefore, matches should be put away, and check the stove to make sure he did not leave it on. Sharp knives should be put in a safe place. He should also avoid driving.

Diagnosis and Treatment:

MMSE scores of 10-20 are considered moderate Alzheimer's disease. MRI and CT scans of the brain are either normal or may show atrophy (shrinking of the brain). EEG shows slowing. Treatment options remain the same, as in the mild form of the disease. Anxiety, lack of sleep, and depression should be treated, but avoid using sedatives, including over-the-8 counter sleep aids such as Benadryl or Tylenol PM. The patient's independence and autonomy should be considered. He may need help in the activities of daily living and in social situations. Safety of the patient should be a very important factor, and the caregiver should provide as safe an environment as possible. One thing to remember is that accidents do happen, so do not blame yourself for any unavoidable problems that arise. See your doctor on a regular basis and read about medications that are available. Consider Aricept (donepezil), Exelon (rivastigmine), Reminyl (galantamine), Namenda (memantine), Vitamin E, Motrin (NSAIDS), and the newer antidepressants.

Severe Dementia:

In this stage, the patient continues to lose the ability to take care of himself. This happens several years after the initial diagnosis. He can no longer function independently. His intellectual functions are severely deteriorated. He may not even recognize his spouse with whom he has lived with for many years. Language and social functions are poor, and he may not be able to comprehend anything. He manifests anxiety and agitation. He needs help with self-care and may develop incontinence of urine and stool. He depends on his caregiver for total care, including feeding, cleaning, bathing, and using the toilet. His limbs are rigid, bent, and flexed. This stage virtually robs him of all intellectual functions. He needs supervision at all times and should not be driving.

MMSE scores of less than 10 are considered severe Alzheimer's disease. An EEG shows diffuse slowing. MRI and CT scans show severe atrophy. PET and SPECT scans show decreased metabolism and blood flow to parts of the brain.

In this stage, symptomatic treatment is very important. A study published in the Journal of the American Medical Association (JAMA), concludes that combining Namenda (Memantine HCI) with Aricept (Donepezil), a commonly prescribed Alzheimer's drug; provides greater cognitive, functional, global, and behavioral benefits to people with moderate-to-severe Alzheimer's disease than treatment with Aricept (donepezil) alone. The results of this study mark the first time that positive results have been observed when using a combination of two drugs for the treatment of Alzheimer's disease. Doctors may prescribe some medications for agitation and delusions. Memantine (Namenda), that is an NMDA antagonist, has shown promise in the later stages of the disease even when prescribed alone. Life support measures and resuscitation issues should be discussed with other family members and healthcare professionals. The patient may need nursing home care. The caregiver needs education and support along with psychosocial, medical, financial, and legal help. Consider hospice.

Terminal Phase

The patient is now bedridden. He may be living in his home or a nursing home. He is unable to care for himself or any of his needs. He is unaware of his surroundings, and needs total care. He may develop several medical problems common to a bedridden and immobile patient. Symptomatic care is the only treatment available. He is very appropriate for hospice care.

Pathology

Old age alone never causes dementia. There is some forgetfulness during old age but nowhere close to what is described here. **We do not know the cause of the disease**. We believe that the outer layer of the brain starts to show some degenerative changes and there is a gradual loss of neurons. Advancing age causes abnormal protein deposit and possible neurofibrillary tangles in the brain. This abnormal protein structure, which are called amyloid plaques, along with neurofibrillary tangles, deposit in the brain. There is either a malfunction or loss of neurotransmitters which leads to the lack of communication between the nerve cells. There are several neurotransmitters which are altered, but acetylcholine remains at the top of the list. Other neurotransmitters that could be responsible are Serotonin, Norepinephrine, GABA, Dopamine, NMDA, free radicals, mutation of messenger RNA, etc.

There is no substantial evidence that toxins cause dementia. **The evidence of aluminum, zinc, mercury and selenium causing Alzheimer's disease has not been substantiated**. There are theories that it may be caused by viral and genetic factors, but the evidence is not overwhelming. There are many people who develop the disease with no family history, indicating that inheritance is not the only mechanism; it may be one of several factors. Some genetic markers of the disease have been identified, but some people with the markers do not get the disease and vice versa. Other factors which may affect the brain are: the immune system, environmental toxins, viruses, genetics, and the big "unknown mechanism."

According to current research, the E-4 polymorphism of Apo lipoprotein E on chromosome 19 increases the risk for getting the disease. Siblings of AD patients have twice the risk, and dementia may start at an earlier age. Less than 1-5% of AD is familial in nature. Other risk factors which have not been evaluated fully, and may cause dementia include, but are not limited to: traumatic head injury, Down's syndrome, lower education level, and an older age of parents. Severe head injury resulting in the loss of consciousness could contribute to Alzheimer's (AD), but more research is needed.

As stated previously, there is no known cause for Alzheimer's disease but there are certain preventive measures that can help. Controlling your blood pressure and blood sugar is vital. There was a recent study which contained more than 1,000 people in Japan that found 27% of those with diabetes developed dementia, compared to 20% of people with normal blood sugar levels. Reducing weight and cholesterol may prove helpful. Eating foods rich in Vitamin E, C, B, Folic Acid, Niacin, and Omega 3 Fatty Acids have been known to have positive effects.

<u>Diagnosis and Treatment:</u>

Testing to predict who is or is not going to develop the disease is still in its infancy because specificity is very poor, 50% never developed the disease. There are severe moral and ethical reasons not to do that type of testing until the time comes. Occasional forgetting, even in the presence of family history, does not mean that one will have the disease. The E-4 polymorphism of Apo lipoprotein E on chromosome 19 increases the risk for getting the disease. Siblings of AD patients have twice the risk. There is less than 1-5% of AD that is familial in nature.

Once Alzheimer's disease is suspected a physician who specializes in Alzheimer's disease (AD) or who has interest in AD (i.e. neurologist, neuropsychiatrist, or psychiatrist) should evaluate the patient. The doctor will examine the patient and do mental status testing, or neuropsych testing. He will look for any treatable cause for the dementia. I recommend early diagnosis with early and persistent treatment with one of the 14 above medicines such as Aricept, Namenda, or Exelon. Early and persistent treatment in the patient will delay nursing home placement.

After the examination, the physician will order some tests. The tests that should be done are: cell count, chemistry panel, thyroid function tests, urinalysis, folate and B-12 level, Chest x-ray, EKG, EEG and CT scan or MRI of the brain may be helpful. Most physicians think that a CT scan has a low yield, but if the patient has a sudden onset of the symptoms, gait disturbances and urinary incontinence early on, or stroke-like symptoms, it is an essential part of the workup. There are some sophisticated procedures such as MRI scan of the brain, PET scan, SPECT scan and spinal tap, which should be done if indicated. A brain biopsy is indicated for brain tumor or abscess, but otherwise it is not indicated. If the patient starts showing signs of AD and is progressing faster and/or he has one-sided weakness or sudden severe changes in mental functioning, further studies might be considered.

After the tests are done, make an appointment to discuss the results. Take a friend or family member with you so you can learn more as to what is happening with the patient. Then you should have a family conference and discuss the findings. If you are not satisfied, get another physician's opinion.

As stated earlier, the cause of Alzheimer's disease is not known. There is no cure, but there are treatments available. There is a lot of research underway, but at this time, we cannot advise how to prevent the disease based on the knowledge we have. Read about Aricept (donepezil), Exelon (rivastigmine), Reminyl (galantamine) or Namenda (memantine). These medications do not change the course of the disease, but they ease mild-to-moderate and sometimes severe symptoms. They prevent the breakdown of acetylcholine, a chemical vital for nerve cells to transmit impulses and communicate with one another. They prevent the patient from getting worse, and it may slow down the disease progression. Other medications used, include Aspirin, Vitamin E, Indocin, and Motrin. The newer antidepressants that raise serotonin (as well as norepinephrine) may help memory problems. It is not proven, but they do help depression, which is 20% in dementia.

One can use memory aids (visit www.NEUROBICS.com) to improve memory and prevent it from getting worse. Some of my patients are able to recognize people are able to name them and show improvement in other cognitive functions. Currently, their symptoms are not worsening and remain unchanged. Aricept (donepezil), Exelon (rivastigmine), Reminyl (galantamine) all increase the neurotransmitter acetylcholine, thereby showing some improvement. Namenda (memantine) acts as NMDA antagonist and can be used in moderate-to-severe AD. If your loved one is on Cognex, your doctor will check liver functions frequently. This is not a requirement with Aricept, or any other medications which I have mentioned above. The dropout rate is lowest with Namenda, and lower with Aricept as compared to other medications. Aricept, Namenda and other medications have proven benefits in cognitive, functional, and behavioral symptoms. A new study suggests that memantine reduces caregiving time by more than 50% per month compared to a placebo. Chelating agents, ginkgo biloba and melatonin have not been studied extensively and have not been proven to be effective.

Some of the treatable causes are medication effect, drugs, alcohol, polypharmacy (multiple medications), infection, diabetes, hypothyroidism, hyperthyroidism, and vitamin deficiencies such as B-1, B-12 and folate. Hypertension and diabetes may cause strokes leading to dementia. One could reduce 50% of vascular dementia by controlling blood pressure. In younger patients, one needs to think about drugs, AIDS, and syphilis, which could cause dementia. In endemic areas, Lyme's disease should be suspected. Normal pressure hydrocephalus presents itself as dementia (memory loss with incontinence and gait disturbance, i.e. magnetic gait, which is to walk as if your feet are attached to floor). Simple shunt placement improves normal pressure hydrocephalus and reverses or prevents progression of the disease.

Aricept: (donepezil) 5 mg and 10 mg; start 5 mg/d and increase to 10 mg/d in one month. Early and persistent treatment. Helps to delay nursing home placement. Better tolerated than other medications.
Eisai/Pfizer
(888) 274-2378 16
www.aricept.com

Exelon: (rivastigmine) 1.5, 3, 4.5, 6 mg. Start 1.5 mg twice a day and increase every two weeks to 6 mg twice a day.
Novartis
(888) 669-6682
www.exelon.com/index.jsp

Reminyl: 4, 8, 12 mg tablets, 4 mg/ml. Start 4 mg twice a day. Increase to 4 mg twice a day every four weeks to 12 mg twice a day.
Janssen
(800) 526-7736
www.reminyl.com

Namenda: (memantine) supplied as 5, 10 mg. Starts 5 mg/d and increase by 5 mg every week to 10 mg twice a day. Forest Laboratory.

(877) 262-6363

www.namenda.com

Avoid: Elavil and medication which could make him confused including, but not limited to, over-the-counter sleeping aids (Benadryl). Avoid sedatives and narcotics if not needed. Discuss this with your doctor.

Vitamin E 400 IU once a day with vitamin C 500 mg twice a day unless taking blood thinners. Ask your doctors. Always take vitamin C with E.

Vitamin B: B-total (tablet or liquid): 1 dose sublingual or oral once a day.

For Agitation: Neurontin, Depakote and Tegretol, if does not help, then try Trazodone, Haldol, Risperdal, Seroquel, and Zyprexa.

For Insomnia: Sleep hygiene, Rozerem, Neurontin, maybe chloral hydrate, Ambien, Sonata, and Haldol.

The Caregiver

You, as a caregiver, are everything to the patient--the servant, cook, wife, nurse, friend, etc. Almost over 15 million people provide unpaid care for demented patient. Educate yourself with information from the Alzheimer's Association and other resources to become knowledgeable about the disease. Take care of yourself, recognize your needs as well, and keep yourself healthy so you can provide good care for the patient.

It is very stressful to take care of a patient with Alzheimer's disease. Staying healthy emotionally and physically is essential. That is why annual physical and mental exams are important for you. Make sure you have a well-balanced diet, as well as getting regular physical exercise with enough rest. If stress becomes a problem, stress-reducing exercises or talking to a family member, friend, or religious member may help. If it continues to get worse, see your physician and/or a specialist.

Going to church and church meetings will reduce anxiety and tension. Some of my patients felt better when attending church services on a regular basis. You may meet some nice people there who can help you. Grief, anger, frustration, depression, anxiety and tension are normal while you are dealing with a patient who has a progressive decline in his memory. Seek counseling and/or join an Alzheimer's group in your area.

All caregivers should have a list of names, addresses, and telephone numbers, including emergency telephone numbers such as the doctors, poison control, fire department, and hospital. The medication schedule should also be written clearly on a notepad with directions on administering them. If someone is babysitting, leave important messages which explain the patient's condition, activity level, medication schedule, diet, likes and dislikes.

You will need to have a written daily schedule, something the patient should be oriented to. Start every day fresh. Orient the patient about the day, date, month, and year every day. Talk to him every day about the schedule and the activities you have planned. Ask someone to sit with him while you do some chores in the house or outside, or go for groceries, etc.

Adult daycare centers are another option. Respite may help in this regard. Take the patient to the senior citizens center. Let him have lunch there, mingle with others and have fun.

Comfort, Hygiene, Abilities

The patient should continue socializing if desired, under supervision. Sensory stimulation needs to continue. We may also need to improve the patient's general physical condition and sleep hygiene.

If you want to give him a bath, watch him carefully while he bathes. If that is not a workable situation, clean the patient with a wet towel. Keep his nails short. The Alzheimer's patient who is capable should have his independence and not be restricted. Let him take care of himself as much as he is able and help him only if necessary. Let him bathe, shower, brush his teeth, eat, pick out his clothes and dress himself according to his abilities.

He will function better with a routine specifically designed for his needs and impairment level. Complicated and multi-step procedures (bathing and dressing) should be broken down into small and simple tasks. For example, to comb his hair, find the comb, hold it in one hand, comb the hair, etc. Make sure you understand the patient and be aware of his needs including hunger, thirst, constipation, incontinence, etc.

For constipation, he should drink plenty of water. At least eight glasses of water a day is a good idea. Stool softeners, prune juice, etc., may help. Bowel incontinence usually happens in the later stages of the disease, but make sure his medications are not causing the problem. Let the doctors check him to rule out other causes of bowel incontinence. Urinary incontinence can be taken care of by regularly taking him to the bathroom and sitting him on the toilet. Pads or a catheter may help, but the doctor will have to decide about the catheter. Wash his privates well to avoid infection. The procedure should be kept simple and easy.

The Alzheimer's patient will lose things. Keep an extra pair of glasses, preferably made of plastic. Keep an extra set of keys. He will hide things and blame it on the caregiver or some unknown person. He will hallucinate and think there are people in the house who are taking his belongings. He will hide things under the bed or couch, behind the dressers, etc. Remember, he is not deliberately forgetting things, so don't get upset or angry when you explain things to him.

Meals

Prepare meals that are easily chewable. Prepare soups, soft diet or liquid diet, depending on his capabilities. Food should be lukewarm. Do not give him a knife. Cut his meat for him. Keep a bib on him and make sure his dentures are in unless they cannot be kept in. Do not feed him if he is too sleepy, angry, or uncooperative. If supervised dinners at restaurants are not possible, try bringing food home and/or cook a nice meal at home. If eating together is not possible, it is okay; do not get upset; feed him first and then have your meal.

Expectations

Depression can be a problem but can be treated easily with newer antidepressants. Depression effects 20-30% of demented patients. In 1960s **depression was termed as Pseudo-dementia. Delusions** (false beliefs i.e. stealing, wife interested in somebody else, etc.) are common, but reassurance will help. If not, newer antipsychotic medication will help. Illusions (misinterpretations) such as a shoe looks like a dog, etc., and hallucinations (see, hear, and feel things that are not there) can be treated if needed again with the newer antipsychotic.

People with Alzheimer's disease get restless, wander off, or get paranoid, more so toward the evening **(sundowning)**. Play music or watch TV programs he likes. If he gets severely agitated, the doctor may put him on medication. When the time comes to go to bed, do not feel bad if you have to sleep in separate beds and/or separate rooms. Keep the room and bed cozy and comfortable. Stimuli such as loud music and TV should be kept to a minimum.

Praise the patient if he understands and takes commands. Speak clearly and loud enough so he can hear you. Repeat yourself only if needed. Do not get angry if you have to repeat the same thing over and over again. Avoid things which will upset him. Remain calm and avoid confrontation if he is very upset. Keep your requests and demands as simple as possible. Write in large letters so he can read what you said if he cannot hear you. Items in the kitchen and bathroom should be labeled with large print making it easier for him to read. Instead of fighting or resisting, redirect his worries. Give him simple activities to keep him busy.

You don't have to worry about all of the problems, but try to take care of one problem at a time intelligently. Continue to be the same kind, loving, and touching person that you were. Do not give up the warmth and the love you had for him.

Safety

Make sure the patient is safe in his home environment. Make some changes in the house for safety, but limit the changes to avoid confusing him, which may make him feel disoriented and displaced. In moderate-tosevere stages of the disease, changes in the house are necessary and sometimes life-saving. There should be handrails in the tub as well as close to the toilet, so he can pull himself up easily.

Keep a first-aid kit at home and in the car for emergencies. Pills, medication, and cleaning supplies should be locked up. Remove small and dangerous objects such as marbles, bottle caps, pellets, etc., that can be swallowed. Weapons should be removed from the home, as they can be dangerous. Floors should be kept clean and non-slippery. Small throw rugs and/or other objects should be removed. Small furniture pieces should be put away. Padding should be placed on the corners of sharp tables. The stairs should have railings and be clearly marked. Cover the stove knobs and cover the electrical outlets. Matches and lighters should be put away. The smoke detectors should always be in working condition. The lighting in the house should be well lit, leave the night-light or bathroom light on. The outside doors should remain locked.

The home should be made as safe as possible. Remember what you did for your toddlers and do the same. Get a cane, walker or wheelchair making sure the wheelchair is light enough for you to carry. The patient's shoes should be comfortable with non-slippery soles. Velcro or slip-on shoes are safer than shoes with shoelaces. If you take him outside, make sure safety precautions are in effect. Supervision should be continued outside as well. Swimming pool areas should be enclosed and locked at all times. It is preferable that he does not drive, although he will not comply. If you drive with him, keep your seatbelt on and doors locked. In the later stages of the disease, use the child safety lock. If he continues to argue, the keys should not be released. The safety of others, as well as the patient should be considered.

Resources for Caregivers

There are many people and organizations to help you. First and foremost are the friends and family members. Ask family members and friends, or a person you trust, to sit with him if you have to go out.

As well as taking the medication list to the doctors, take one to the daycare center also and give it to the personnel. A daycare center will help him socialize and help him remain active, which will help him sleep better and build his self-esteem. It provides a break for the caregiver. Daycare centers may cost you about $30 a day or so.

If family and friends cannot help, a live-in aide or a person who can stay with the patient for a few hours will be necessary. Look for support groups such as the Alzheimer's Association and places where professionals will help. Attend support groups and learn how others are taking care of similar situations. It does not mean that you are giving up; it means that you are actively seeking information on coping skills. Daycare centers, nursing homes, respite, registered nurses, home health aides, and social workers can help. Visit a nursing home to see how good the care is and make a decision depending on the patient's medical condition and your financial status.

Finances and Legal Help

Alzheimer's disease is one of the most expensive diseases to take care of. Start gathering information from the Alzheimer's Association, counselors, social workers, attorneys, etc., in your area.

One of the most important items is to talk to his spouse, or if you are the spouse, talk to family members. Gather information about insurance policies, income, assets, medical care and other expenses such as home healthcare, home-sitters, daycare centers, nursing homes, etc. He should have a Power of Attorney and Living Will in case a decision needs to be made regarding his care. You should start gathering financial records for the previous three to four years to find out his average monthly/yearly expenses and income. Obtain information about bills, medication charges, and medical expenses. After you have collected all of the pertinent information, then you can talk to an attorney or an accountant.

There are several expenses the patient will incur in addition to medications, such as accountant, attorney, home-sitters, visiting nurses, meals, transportation, telephone, electricity, making the house safe, remodeling, long-term care if needed, etc.

All bank account statements, bank books, canceled checks, wills, trusts, insurance policies, retirement policies, business and legal papers, tax returns, and bills showing expenses must be tracked. If bills have not been paid, request that the company contact you directly, so you can make payment arrangements. The post office should send the bills directly to you if you are not living with the patient.

Keep a record of insurance numbers and his Social Security number. You can write these numbers, important dates, and his date of birth in one corner of the medication list.

All stocks, bonds, safety deposits, and life insurance policies should be taken to an accountant and/or attorney. (Note: You cannot get into the safe deposit box without a court order unless you are the patient's spouse.) Check into his real estate holdings and have a CPA and/or attorney advise you of tax benefits and taxes owed.

If you need long-term care insurance, try different agencies and apply for a policy. If the patient is under 65 and has been disabled for two years, he qualifies for Medicare. If he is older than 65, he is eligible for Medicare anyway, and it should be applied for. Medicare may cover a home health care agency if the physician prescribes it. Medicare will pay for nursing home care according to their guidelines.

Medicaid is also available. It is a state-funded insurance program. You will need to contact an expert on the laws and the benefits. The laws are difficult to understand and do not cover custodial care. You will need to ask an attorney or social worker who understands insurance to help you apply for benefits. If you apply for Medicaid, be sure that gifts made within the last three years of the application are not waived. His home is exempt and the gifts are exempt, but contact an accountant and/or attorney who specializes in, and is familiar with the law. The Medicare and Medicaid programs are extremely technical and are constantly changing.

The patient should have a durable Power of Attorney, which allows a person to act as an agent in case he is unable to make his own decisions. Living Trusts are good for property management, which allows the trustee of the living trust to manage the property. Living Wills or a Health Care Proxy will guide the doctors or healthcare professionals on how to proceed in case of severe illness from which the hope for recovery is not reasonable. Health Care Proxy gives a person permission to make decisions in the event the patient cannot make his own decisions. Health Care Proxy can be obtained by an attorney or the organization Choice in Dying. All of these issues should be discussed with a social worker, attorney, and accountant so the best decision can be made for the patient.

Types of Dementias:

There are several causes of dementia but we think that AD is underlying disorder in most if the patients. They are estimated to be more than 100.

***Alzheimer's Dementia**: most common form. More than 60% of all dementias is due to AD. No known cause. It could be mild, moderate or severe. Or 7 stages in which stage 1 is normal, stage 7 is severe.

***Vascular Dementias**: second most common. It is due to lack of blood flow to brain due to strokes or small blood vessels plugging up. It is aggravated with small strokes, poorly control high blood pressure, poorly control diabetes. Lack of circulation to brain is the major cause if this type of dementia.

***Dementia with Lewy Bodies:** also called Lewy Body dementia. An abnormal protein deposits in brain cell called Lewy bodies which causes brain cells to function abnormally. It causes dementia, tremors and Parkinsonian features.

***Parkinson's Dementia Complex:** In advance Parkinson's disease the cognitive functions decline leading to dementia. Not all patients with Parkinson's disease gets dementia.

***Frontotemporal Dementia**: Called Picks disease. It involves frontal and temporal lobes. It affects personality, social skills and behavioral changes. Following which they developed memory and speech problems.

***CJD (Creutzfeldt-Jacobs Disease):** also known as Mad Cow disease. Rare disorder afflicts 1 in one million. Caused by a virus. It is a rapidly progressive disorder over several months. Rapidly progressive forgetfulness, muscle jerking, twitching, muscle stiffness, poor coordination, hallucinations and visual problems etc.

***NPH (Normal Pressure Hydrocephalus)** increased fluid in the brain ventricles. That will compress the brain against the skull. It causes memory problems, incontinence of urine and wide based magnetic gait. They walked with wide based gait as if feet are attached to the ground.

***Dementia related to Huntingston Chorea** : dominantly inherted. Cognitive, movements and behavioral disorders. Poor judgement, mood swings, speech involvement, delusions, hallucinations, jerking of extremities. Uncontrollable jerking of different body parts.

***Sometimes dementia is caused by more than one medical condition. This is called **mixed dementia**. The most common form of mixed dementia is caused by both Alzheimer's and vascular disease.

*****Wenicke-Korsakoff Syndrome:** related alcoholic overuse, Thiamine deficiency, visual and other eye problems, trouble walking, continue to drink and gets worse till slips into coma. Amnesia of events, false perception of things, hallucinations, malnourished chronic alcoholic. Treatment is stop drinking, taking Vitamin B-1 Thiamine 100 mg/d.

***<u>**MCI (Mild Cognitive impairment)**</u>: it is the stage between normal aging and dementia. The problems with memory, thinking, language and judgment are worse than normal. It may not affect the daily activities. The risk of dementia is increased.
Forgetting more, losing train of thought in conversations, forgetting appointments, poor decision making, impulsiveness, irritability, depression sets in. brain changes are similar to AD but milder in nature. 10% of older patients develop dementia every year.

<u>What to do for an appointment with doctors:</u>

Prepare a few days before the appointment. Write down all your symptoms. Be brief. Write down all your medical problems. Make a list of all medical conditions. Make a list of all your medications. Try to take any test results you have with you to an appointment. Do not assume that the doctor will have all the records. Take a friend or family member with you.

LIST OF RESOURCES:

Alzheimer's Sites:

www.aoa.gov

www.eldercare.gov

www.alzheimers.org

www.alz.org

Alzheimer's Association

For Alzheimer's disease and related disorders.

Provides educational material, resources, support group locations, help organizing a support group if one is not available.

(312) 335-8700

(800) 272-3900

www.alz.org

Alzheimer's Disease Education and Referral Center

Helps educate families and healthcare professionals regarding the disease.

(800) 438-4380

www.alzheimers.org

National Academy of Elder Law Attorneys

Assists attorneys, bar organizations and others who work with older clients and their families.

(703) 942-5711

www.naela.com

Administration on Aging

Provides a comprehensive overview of a wide variety of topics, programs and services related to aging.

(202) 619-0724

http://www.aoa.gov 25

National Institute on Aging

Conducts and supports research, training and disseminates research findings and health information on aging processes, diseases and other special problems, and needs of older people.

(301) 496-1752

http://www.nia.nih.gov/

National Rehabilitation Information Center

Information services on disability and rehabilitation issues. Articles, books and literature.

(800) 346-2742

http://www.naric.com

Center for Medical Consumer and Healthcare

An advocacy organization to improve the quality of health care supported by private contributions, newsletter subscriptions and the Judson Memorial Church.

(212) 674-7105

http://www.medicalconsumers.org

Accent on Information

A newsletter with resources available online and information for the handicapped.

(800) 834-3059

http://www.accentcare.com

Contact Helpline

A 24-hour, 7 day a week, listening, information, and referral service.

(717) 652-4400

National Hospice Organization

Provides quality, compassionate care for people facing a life-limiting illness or injury.

(703) 837-1500,

http://www.nho.org

Family Caregiver's Alliance

Supports and assists caregivers of brain-impaired adults through education, research, services and advocacy including an online support group for friends and family members of an adult with cognitive disabilities.

(415) 434-3388

(800) 445-8106

www.caregiver.org

Elder Care Locator

Connects older Americans and their caregivers with sources of information on senior services, provides links to community-based organizations that serve older adults and their caregivers.

(800) 677-1116

http://www.eldercare.gov

National Mental Health Association

Works to improve the mental health of all Americans with mental disorders through advocacy, education, research and service.

(703) 684-7722

(800) 969-NMHA (6642)

http://www.nmha.org/

Suicide Prevention

(310) 391-1253 nationwide 1-800-suicide

(800) 784-2433

Medic Alert

Jewelry such as bracelets and necklaces on which health alerts are imprinted.

(888) 633-4298

Aricept (888) 274-2378

Consumer information for Aricept 888-422-4743

Cognex (800) 223-0432 or (303) 629-9384 after hours.

The Caregiver's Marketplace

Offers products to help with daily activities such as eating utensils, lipped plates, magnifiers, and bath aids. Catalogs and online ordering are available.

(800) 323-5547

http://www.caregiversmarketplace.com

Thanks

Dr. Kazmi.

ALZHEIMER'S IS TREATABLE

Alzheimer's disease is now a treatable disorder. You can do few things to prevent it from getting worse or delay it or slow it down.

Facts

*A FATAL DISORDER, 6 TH LEADING CAUSE OF DEATH

*UNKNOWN CAUSE

* ONE IN 8 ABOVE 65 YRS. HAVE IN USA

* 5.4 MILLION HAVE ALZHEIMER'S IN USA

*15 MILLION PEOPLE PROVIDE UNPAID CARE IN USA

*2/3 ARE WOMEN

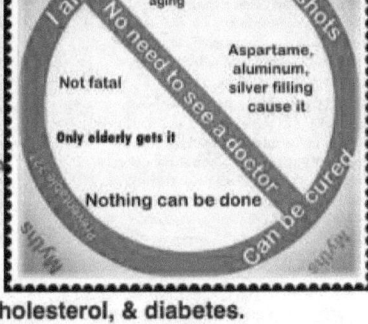

Lifestyle Changes

- Control your weight, blood pressure, cholesterol, & diabetes.
- Keep your mind active with daily physical, mental, & spiritual exercises.
- Learn to do relaxation techniques and deep breathing exercises.
- Socialize : maintain friendships & social skills, do not isolate
- Retirement will retire your body, soul & brain.
- Walk briskly 7-12 thousand steps per day, or a minimum of 30 minutes.
- Stimulate the non-dominant side of your brain (i.e., use your non-dominant hand).
- "Neurobics" should be done every day (i.e., cards, word puzzles, number games, brain teasers, web: games.AARP.org etc, etc. go on line & check it out.
- No more than 2 alcoholic drinks daily – one is better
- Stop smoking: good for health, especially Alzheimer's.

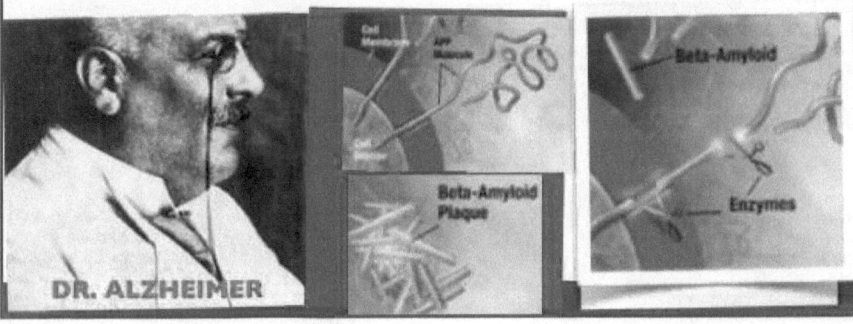

DIET: EAT FOR BRAIN: Remember the "F diet" of Dr. Kazmi which includes fowl, fish, fruits, fresh vegetables, & fresh clean water.

Carbs
A Low-carbohydrate diet is associated with a 30% reduction in Alzheimer's.
Yellow pigment in curry breaks beta amyloid plaques; People who eat curry often (more than once a month) performed significantly better in mental tests. Eat more fruits, vegetables, and berries.
Eat a well-balanced diet which is low in carbs and saturated fats.

- **Good fats:** MUFAs & PUFAs, liquid at room temperature.
  - Monounsaturated Fats MUFAs : lower bad cholesterol (LDL), and raise good cholesterol (HDL)-They may also improve insulin levels and sensitivity, especially for type 2 diabetics. Nuts, avocados, olive oil
  - Polyunsaturated Fats: PUFAs : any plant based fats, they contain Omega 3 fatty acids. Sea food: salmon, fish oil; corn, soy, safflower, sun flower oil, canola, flax seed oil

- **Bad fats:**
  - saturated fats: animal products, dairy, egg, palm oil, palm kernel oil,
  - Transfat: hydrogenated liquid oil synthetically processed to withstand cooking, found in packaged food, fast food

- **Others**
  - Increase intake of nuts, seeds, avocados, poultry. Omega 3 Fatty Acid Containing foods include: Egg whites, nuts, walnuts, green leafy vegetables. Cold water fish: salmon (not from a fish-farm), albacore tuna, mackerel, herring, sardines, lake trout.
  - B6 and B12 containing food: Bananas, beans, chicken breasts (B6), shellfish, salmon, trout, liver, and lean beef (B12).

- **Antioxidants**
  Antioxidants protect the brain from free radicals which are elevated in Alzheimer's.
  Foods rich in antioxidants include:
  - fruits: blackberries, blueberries;
  - vegetables: artichokes, red cabbage, spinach, broccoli, sweet potatoes; nuts such as walnuts, pecans;
  - spices: turmeric, ground cinnamon.
  - Foods rich in Vitamin E: blackberries, blueberries, spinach, broccoli.
  - Foods rich in beta-carotene: leafy green and yellow vegetables like broccoli, sweet potatoes.

Send comments to Dr. Kazmi, 928-855-8979; neurodoc60@icloud.com

SUPPLEMENTS

Vit B-12,

Vit C

Vit E 400 u/d

Niacin (slow release)

Coconut oil

Multiple vitamin

Pointers:
- **Start treatment of Alzheimer's early, do not stop the treatment.**
- **If homocysteine level is high, use Cerefolin or high doses of Vitamin B- complex**
- **Estrogen does not slow down the disease**
- **Iron deficiency plays a role, check & get treatment**
- **Familial dementia gets worse with low B-12**
- **Yellow pigment in curry breaks down the Beta Amyloid**
- **Eat more fruits vegetables & berries**
- **Join a support group**
- **Educate**
- **See a doctor to start treatment as early as possible.**

Avoid
Over the counter sleep aids.

Depression, Stress, living in Isolation

Carbohydrates, extra Salt/Sugar

Red meat, saturated fats.

Heavy drinking

Inactivity

Do mental, physical, emotional, spiritual, social exercises / activities on a daily basis. Zestful sleep. Reduce stress. Be happy.

Alzheimer's
Disease is Treatable

Delay Alzheimer's Disease ?

- Be motivated, be positive
- Control blood pressure, weight, cholesterol, diabetes, & other health problems
- Consume foods rich in omega 3 fatty acids (fish, flaxseed, soybean)
- Add vitamin B, C, E, folic acid, niacin to your diet
- Add Vitamin B-12 for sure
- Keep your mind active with mental, physical, emotional, social exercises / activities
- Neurobics : brain games, AARP / Web MD, Lumosity, word puzzle, card games, number games
- Relaxation
- Deep slow breathing exercises
- Socialize
- Make friends, do not isolate
- Treat depression, anxiety
- Walk briskly 7-12 K steps per day or 30 minutes of nonstop workout
- Aerobics
- Treat any treatable disease
- Control any preventable diseases
- Sleep well
- Diet: Mediterranean diet
- Dr. Kazmi's diet : "F" fish, fowl, fruits, fresh vegetable, fresh water
- A low carbohydrate diet : 30% reduction of dementia
- Turmeric spice may be helpful
- Good fats : MUFA & PUFA : liquid at room temperature, unsaturated fats.
- MUFAs: nuts, avocados, olive oil
- PUFAs : any plant based fat, contains omega 3 fatty acids, in sea foods: salmon, fish oil, soy, sunflower, canola, flax seed oil
- Bad fats : saturated fats, trans fat
- Bad fats : Saturated: animal products, dairy, egg, palm oil, palm kernel oil
- Bad fats : Trans fat: hydrogenated liquid oil, synthetically processed (package food, fast food)
- Increased intake of nuts, seeds, avocados, poultry, omega 3 fatty acids, egg white, green leafy vegetable, cold water fish : salmon, albacore, tuna, mackerel, herring, sardines, lake trout
- B-6 & B-12 : bananas, beans, chicken breasts (B-6), shellfish, trout, liver, lean beef (B-12)
- Antioxidants : consume more: fruits, blackberries, blueberries, vegetable, artichokes, red cabbage
- Start these as soon as possible
- Start treatment of dementia as soon as possible
- Do not stop treatment because improvement is very slow. Join support group
- Dated : 11-1-2015 By Dr. Kazmi MD. 928-855-8979

www.ingramcontent.com/pod-product-compliance
Lightning Source LLC
Chambersburg PA
CBHW030012190526
45157CB00015B/2459